3/97

SECRETS OF THE ANIMAL WORLD

CRUSTACEANS
Armored Omnivores

by Andreu Llamas
Illustrated by Gabriel Casadevall and Ali Garousi

Gareth Stevens Publishing
MILWAUKEE

For a free color catalog describing Gareth Stevens' list of high-quality books and multimedia programs, call 1-800-542-2595 (USA) or 1-800-461-9120 (Canada). Gareth Stevens Publishing's Fax: (414) 225-0377.
See our catalog, too, on the World Wide Web: http://gsinc.com

The editor would like to extend special thanks to Jan W. Rafert, Curator of Primates and Small Mammals, Milwaukee County Zoo, Milwaukee, Wisconsin, for his kind and professional help with the information in this book.

Library of Congress Cataloging-in-Publication Data

Llamas, Andreu.
    [Cangrejo. English]
    Crustaceans: armored omnivores / by Andreu Llamas; illustrated by Gabriel Casadevall and Ali Garousi.
        p. cm. – (Secrets of the animal world)
    Includes bibliographical references (p. 31) and index.
    Summary: Describes the physical characteristics, habitat, behavior, and different kinds of these armored invertebrates.
    ISBN 0-8368-1585-8 (lib. bdg.)
    1. Crustacea–Juvenile literature. [1. Crustaceans.] I. Casadevall, Gabriel, ill.
II. Garousi, Ali, ill. III. Title. IV. Series.
    QL437.2.L5813   1996
    595.3–dc20   96-17835

This North American edition first published in 1996 by
Gareth Stevens Publishing
1555 North RiverCenter Drive, Suite 201
Milwaukee, Wisconsin 53212 USA

This U.S. edition © 1996 by Gareth Stevens, Inc. Created with original © 1993 Ediciones Este, S.A., Barcelona, Spain. Additional end matter © 1996 by Gareth Stevens, Inc.

Series editor: Patricia Lantier-Sampon
Editorial assistants: Diane Laska, Rita Reitci

Printed in the United States of America

1 2 3 4 5 6 7 8 9 99 98 97 96

CONTENTS

THE CRUSTACEAN'S WORLD

Where crustaceans live

Crabs, lobsters, and shrimp belong to the scientific class Crustacea, which includes thirty thousand different species with "outer skeletons" that protect their bodies like a suit of armor. Although some species live on land, about 99 percent spend their lives in all types of water habitats, such as seas, lakes, and rivers. Some crustacean species have also adapted to life in the currents of underground water sources.

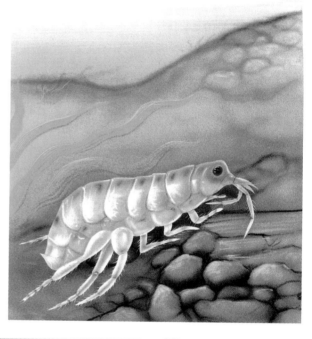

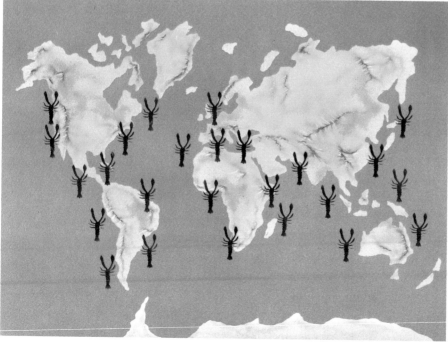

The sand flea is a crustacean that lives in sea water but also jumps around in sand.

Crustaceans are hardy animals that can survive almost anywhere. Most species live in aquatic environments.

4

Animal armor

As a natural defense against enemies, crustaceans have a hard, heavy shell that covers their body and turns them into prey that are difficult to catch and eat. This armor is made of overlapping plates that work much the same way as the armor of medieval knights.

Crustacean armor is several layers of a hornlike substance called chitin that is made by the outer layer of skin, or epidermis. The layers are very hard because they contain a lot of the element calcium; calcium is what makes bones hard.

This crab, protected by its armor and large claws, can be a difficult prey to catch.

Some crustaceans, like these sow bugs, can roll their body up into a ball for extra protection if they sense danger.

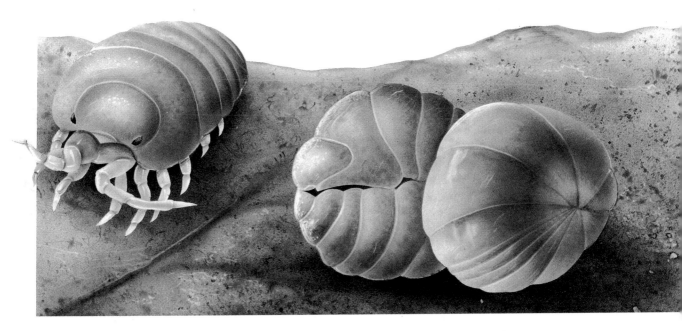

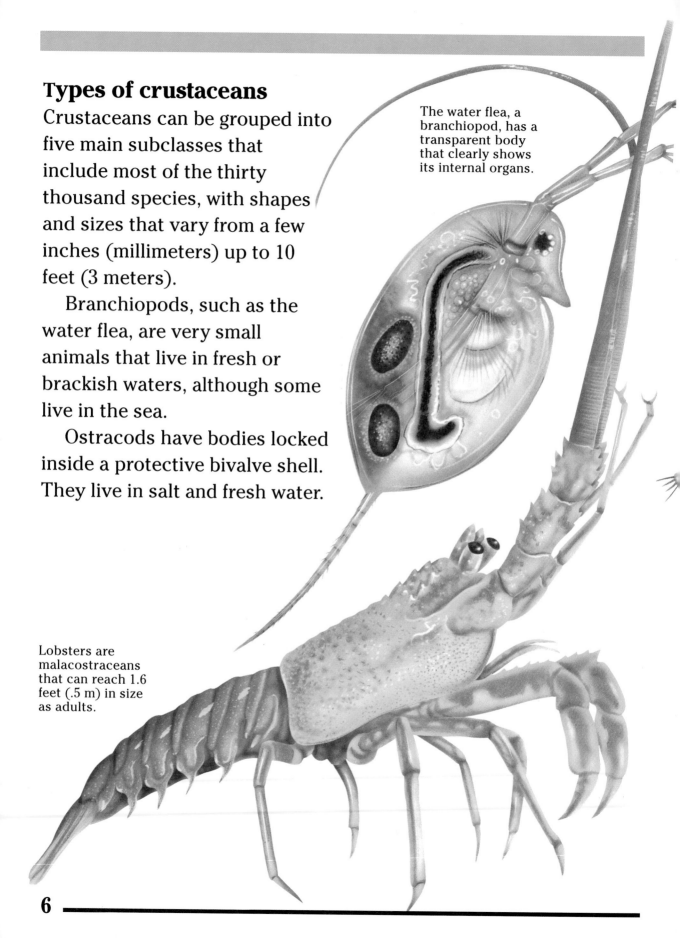

Types of crustaceans

Crustaceans can be grouped into five main subclasses that include most of the thirty thousand species, with shapes and sizes that vary from a few inches (millimeters) up to 10 feet (3 meters).

Branchiopods, such as the water flea, are very small animals that live in fresh or brackish waters, although some live in the sea.

Ostracods have bodies locked inside a protective bivalve shell. They live in salt and fresh water.

The water flea, a branchiopod, has a transparent body that clearly shows its internal organs.

Lobsters are malacostraceans that can reach 1.6 feet (.5 m) in size as adults.

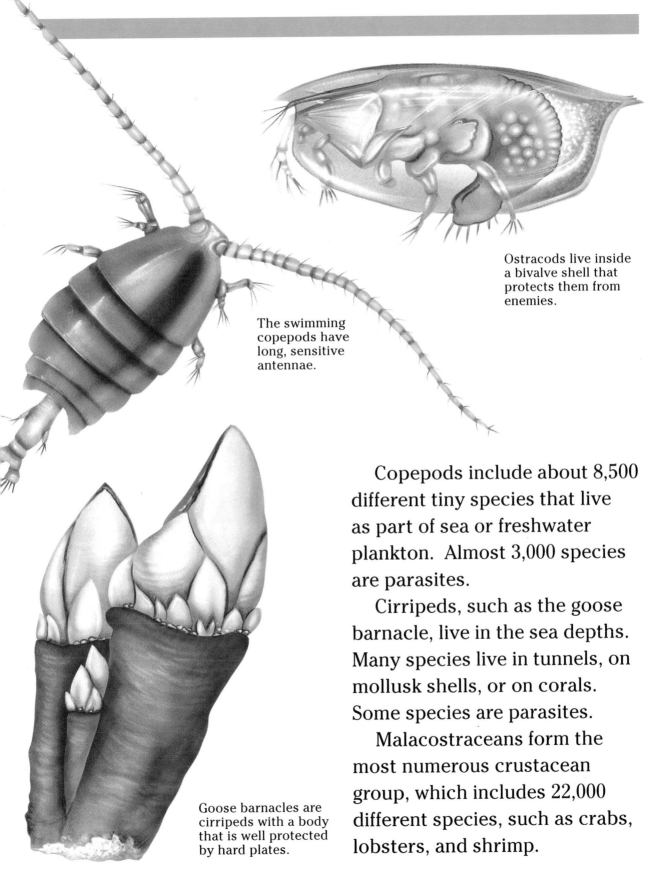

Ostracods live inside a bivalve shell that protects them from enemies.

The swimming copepods have long, sensitive antennae.

Goose barnacles are cirripeds with a body that is well protected by hard plates.

Copepods include about 8,500 different tiny species that live as part of sea or freshwater plankton. Almost 3,000 species are parasites.

Cirripeds, such as the goose barnacle, live in the sea depths. Many species live in tunnels, on mollusk shells, or on corals. Some species are parasites.

Malacostraceans form the most numerous crustacean group, which includes 22,000 different species, such as crabs, lobsters, and shrimp.

INSIDE THE LOBSTER

The lobster's body is easily recognized by the tough outer armor that helps protect it from enemies.

Lobsters feed on small, live animals that they trap with their huge claws. The victims are pushed into the mouth and chewed into small fragments by the buccal pieces. If the lobsters cannot capture live animals, they can also feed on dead plants and animals.

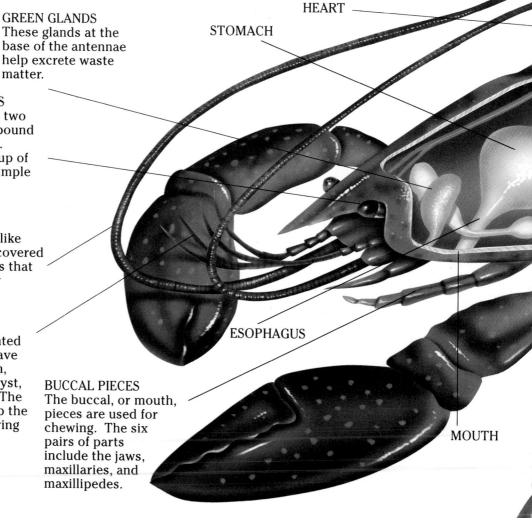

GREEN GLANDS
These glands at the base of the antennae help excrete waste matter.

STALKED EYES
Lobsters have two movable compound eyes on stalks. Each is made up of hundreds of simple eyes.

ANTENNAE
The long whiplike antennae are covered in minute hairs that act as sensory detectors.

FEELERS
These two jointed feelers each have a special organ, called a statocyst, at their base. The statocysts help the lobster in hearing and balance.

BUCCAL PIECES
The buccal, or mouth, pieces are used for chewing. The six pairs of parts include the jaws, maxillaries, and maxillipedes.

HEART

STOMACH

ESOPHAGUS

MOUTH

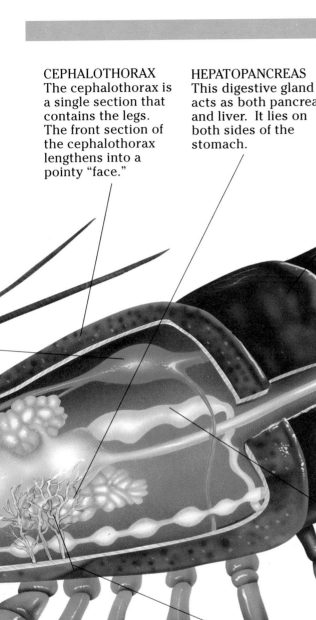

CEPHALOTHORAX
The cephalothorax is a single section that contains the legs. The front section of the cephalothorax lengthens into a pointy "face."

HEPATOPANCREAS
This digestive gland acts as both pancreas and liver. It lies on both sides of the stomach.

ABDOMEN
The flexible abdomen is made up of six rings that can fit inside one another, folding up the abdomen underneath the cephalothorax.

SHELL
The body is protected by a tough, but flexible, shell made of chitin and calcium. The body has three parts: cephalothorax, abdomen, and tail fin.

INTESTINE

TAIL FIN
The tail fin is made up of five segments. When the lobster is attacked, it darts backward by rapidly beating its tail fin.

LOCOMOTION
The lobster walks slowly, using its last four pairs of legs. The first pair, made up of claws, is used to capture food.

GILLS
The lobster breathes through gills, or branchias, that are under the shell. To help it breathe, the animal uses its extremities to create a flow of water toward the gills.

MOLTING CRUSTACEANS

Changing armor

The crustaceans' protective shell has one disadvantage. During the lobster's growth, for example, the hard armor that covers its body does not increase in size along with its internal organs. So at certain times, the lobster needs to molt, or shed, its armor for another, larger one. Because of this, the lobster's overall growth is not steady or regular, but is carried out in periodic molts. These molts are necessary for the shell size to adjust to the growth of

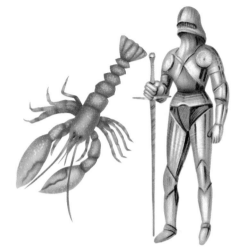

The protective crustacean shell is like a suit of armor.

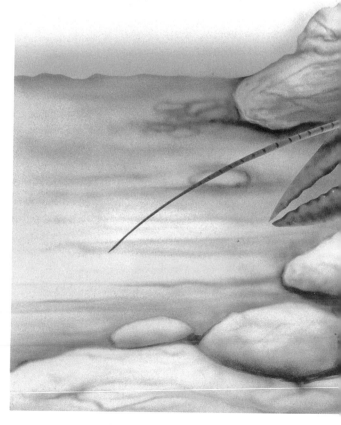

Many crustaceans eat their old shells after molting. This way, they can absorb the calcium.

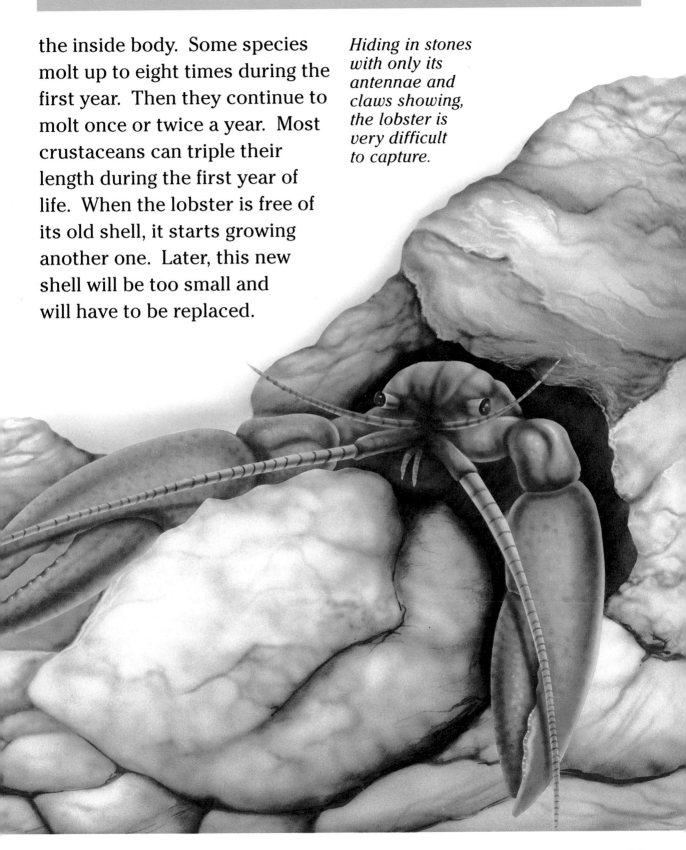

the inside body. Some species molt up to eight times during the first year. Then they continue to molt once or twice a year. Most crustaceans can triple their length during the first year of life. When the lobster is free of its old shell, it starts growing another one. Later, this new shell will be too small and will have to be replaced.

Hiding in stones with only its antennae and claws showing, the lobster is very difficult to capture.

The molting process

Molting begins when the membranes that join the lobster's abdomen with its cephalothorax begin to rip apart. A small tear forms along the back of its shell.

The lobster moves nervously for a few hours. Its sharp, continuous movements help separate the body a little at a time from the inside walls of the shell. First the soft parts contract, and the animal pulls them free from the shell's interior. Then the lobster pushes on the small crack that appears on its shell. The armor separates into two halves, and the animal climbs to the outside.

What seems to be two lobsters is really one that has just crawled out of its old shell.

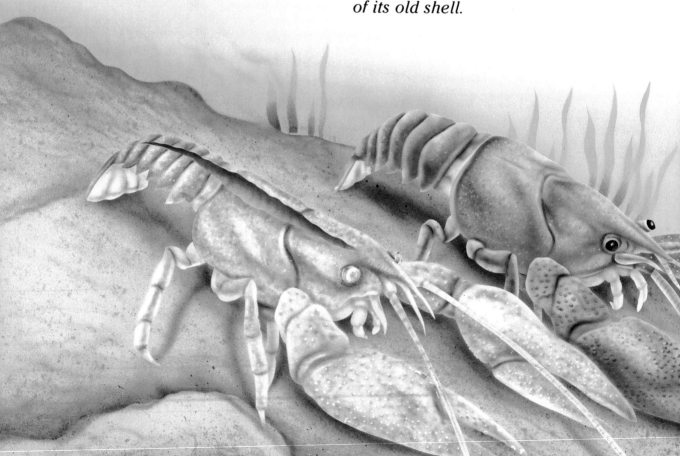

When molting begins, the lobster makes sharp movements to separate its body from the shell's interior walls.

This is a dangerous time in the lobster's life. When a lobster has just emerged from its shell, it is completely helpless to its enemies. During this time, it likes to remain very still, using all its energy to make a new shell with the calcium carbonate it has accumulated in its body during the year.

With great effort, the lobster breaks free of its old armor.

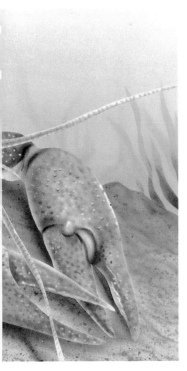

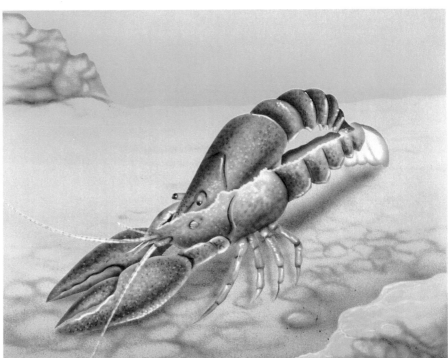

ARMORED DEFENSE

Repairing the armor

During molting, the lobster can sometimes leave one of its limbs behind in the old shell. This is not a serious problem for the animal, since limbs can be regrown. But the regrown limb can sometimes be weaker than before or even deformed.

When a lobster is unable to outrun an enemy that is about to capture it, the crustacean may use an emergency plan: it can voluntarily detach a body part, such as a leg. Then the lobster can escape while the enemy is busy eating the lost limb.

This lobster detaches one of its legs to escape a predator. The limb is devoured while the lobster escapes.

Lobster enemies

Although lobsters have protective armor and powerful claws, they walk very slowly. For this reason, they appear easy prey to many other animals. When the lobster feels threatened, it tries to escape its enemies by suddenly changing direction and speeding away with the help of its tail.

Crayfish enemies include otters, fish, raccoons, ducks, and storks.

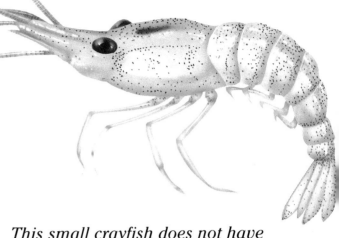

This small crayfish does not have large claws for defense. It relies on speed to escape its enemies.

Coastal crabs also have many enemies that wait for low tide to attack and devour their prey.

The threatening appearance of many crustaceans can scare off some enemies.

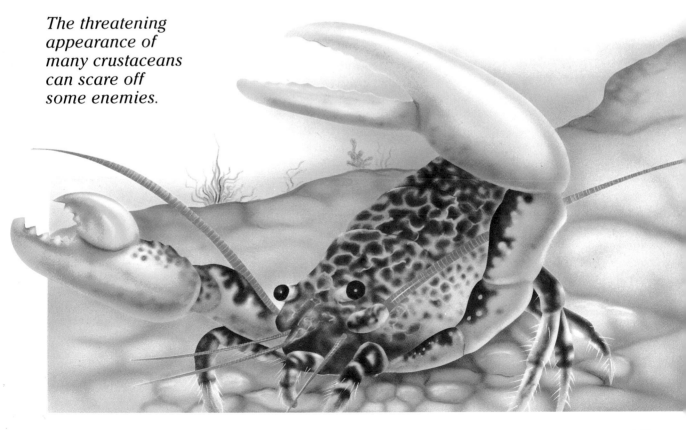

Surprising hermit crabs

The hermit crab has a soft abdomen and no protective shell, so it has adapted itself to living inside the empty shells of other mollusks. Its right claw is larger than the left one. It uses this claw to capture prey and to block its shell entrance. Right after birth, the hermit crab looks for an empty shell that is the right size. Once it finds a satisfactory shell, it moves right in.

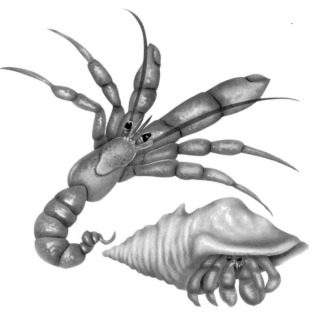

The hermit crab has a hook that it fastens inside the shell it wants to live in.

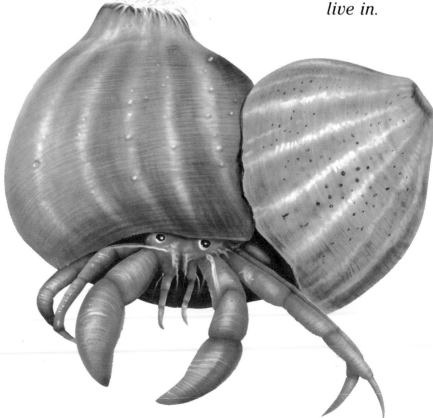

Anemones often travel along on the shell in which the hermit crab lives. Their tentacles help protect the crab from enemies.

that some crustaceans measure more than 10 feet (3 m)?

The largest crustaceans live in the Pacific Ocean close to Japan. The giant Japanese spider crabs can reach up to 12 feet (3.7 m) from the tip of one claw to the other. In ancient times, some fishermen claimed to have seen terrifying Japanese spider crabs 33 feet (10 m) long. If captured, these gigantic crustaceans could feed an entire family. Imagine the excitement of seeing one on the beach while sunbathing!

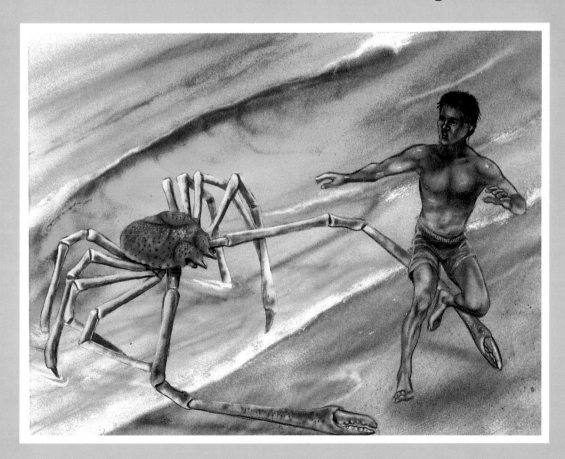

CRUSTACEAN ANCESTORS

The first crustaceans

Crustaceans belong to the scientific phylum, or group, Arthropoda, whose members first appeared 600 million years ago. Early representatives of this group included the trilobites, named because their bodies were protected by a shield divided into three lobes, or sections. Fossils of over 2,000 species have been found, varying between 0.2 and 28 inches (6 mm and 70 cm).

Some trilobite fossils were found folded into balls.

Some trilobite species survived more than 350 million years.

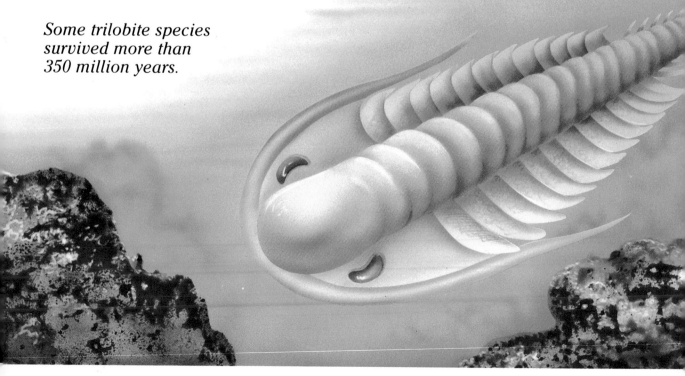

Primitive crabs

Crabs, because of their hard shells, have left abundant, well-preserved fossils. One of the earliest fossils is *Palaepalaemon newberry*, which scientists believe lived in Earth's seas 370 million years ago.

Since the hard parts of the animals preserve best, scientists carefully study the crustacean's shell. This provides them with information that helps them identify the different species.

This crustacean fossil remains well preserved in a chunk of rock.

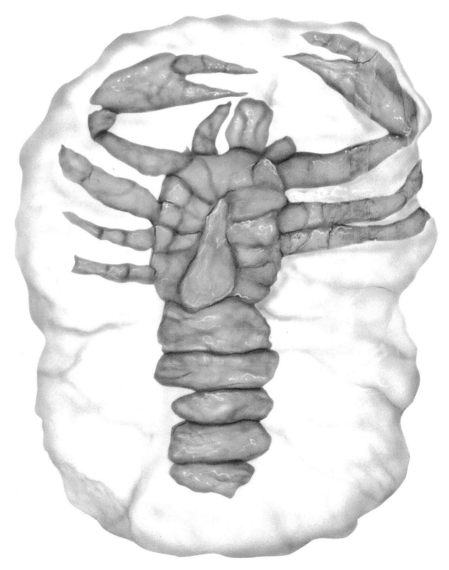

that baby crayfish travel by hanging on to their mother?

After mating, the female crayfish carries 100 to 150 brilliant little eggs attached to her abdomen. This presents a great danger from predators. Female crayfish find shelter when the baby crayfish are ready to start hatching. After birth, the babies hang on to filaments on the mother's belly, and they travel this way for several days. During this time, their soft shells harden, and they learn to swim and walk.

CRUSTACEAN LIFE

Crab senses

A crab's body is covered with sensory hairs through which the animal can smell, taste, and touch. These hairs are more plentiful in the areas close to the antennae, feelers, and buccal pieces.

The eyes, however, are the most specialized sensory organs. The crab has two compound eyes at the end of narrow stalks that can move around and give the crab a wide range of vision.

Some crabs search the beach sand for animal remains to devour, such as the pelican pictured below.

Crustacean compound eyes are made up of small, independent units that vary in number from four to more than ten thousand.

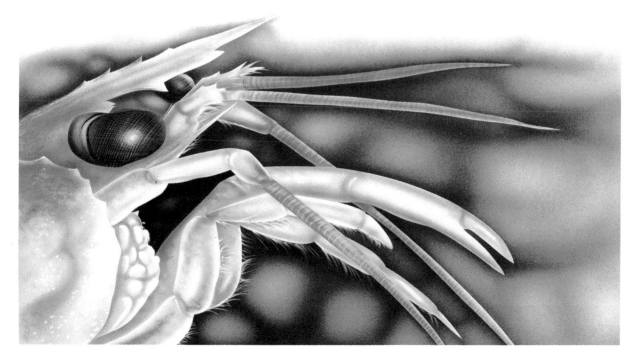

that crustaceans eat almost anything?

Lobsters are omnivorous, like other crustaceans. This means they can eat almost anything they capture, whether plants or animals.

Crayfish come out of their shelters at sunset to look for food. They wait patiently until an unsuspecting prey passes by.

Crabs eat worms, snails, small fish, aquatic reptiles, and amphibians. They also devour the eggs and larvae of other animals and search for decaying remains.

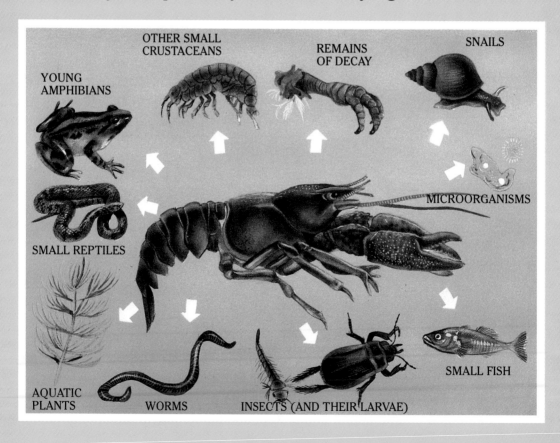

YOUNG AMPHIBIANS

OTHER SMALL CRUSTACEANS

REMAINS OF DECAY

SNAILS

MICROORGANISMS

SMALL REPTILES

AQUATIC PLANTS

WORMS

INSECTS (AND THEIR LARVAE)

SMALL FISH

Dangerous mating ritual

During autumn, crayfish males that are ready to mate chase after females. The females, not having the same urge to reproduce, try to escape. A male that captures a female tries to place her upside-down to mate. If her struggles prevent this, he flings the female against the ground, causing injuries that can be fatal. If the female dies, the male devours her. Once the female is upside-down, the male deposits sperm on her abdomen. The sperm remains attached to

The female crayfish stays in a nest while her eggs incubate.

the female's shell until spring, when she produces eggs. The sperm fertilizes the eggs, which stay attached to the mother's shell until the babies hatch.

This male crayfish is ready to deposit sperm on the female's abdomen.

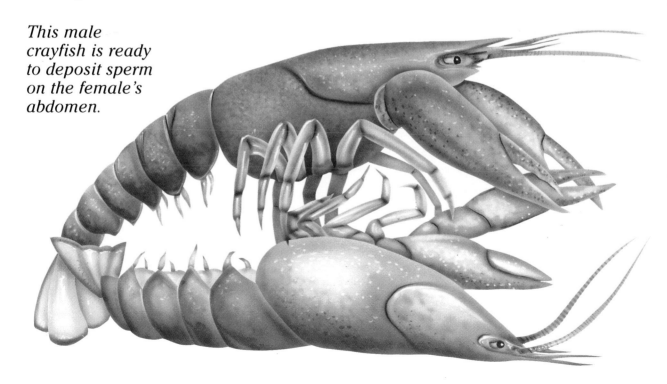

APPENDIX TO

SECRETS OF THE ANIMAL WORLD

CRUSTACEANS
Armored Omnivores

CRUSTACEAN SECRETS

Thousands of eggs. Chinese crabs reproduce easily. The females lay about 500,000 eggs!

▼ **Coconut crab.** The coconut crab lives on islands in the Pacific Ocean and measures up to 18 inches (45 cm) long. It climbs coconut trees and opens coconuts with its powerful claws.

▼ **Shaking off the enemy.** Some crabs are experts at camouflage. They cover their shell with rocks, seaweed, sponges, and other sea life.

▼ **Living buried.** Some species of crustaceans live under the sand. The masked crab, for example, digs in the sand with its legs; during the day, only the

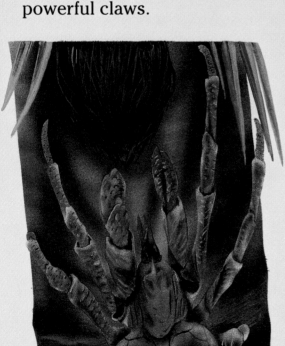

long points of its antennae peek out. To breathe, its antennae have interlaced hairs that work together to make a respiratory, or breathing, tube.

▶ **Cleaning season.** Some species of crustaceans specialize in cleaning the parasites off other animals. Fish travel to areas where these "cleaning stations" are and let the crustaceans pass along their bodies, capturing all the parasites they find. This is a beneficial relationship for both animals, since the fish manage to stay clean, and the crustaceans capture easy prey.

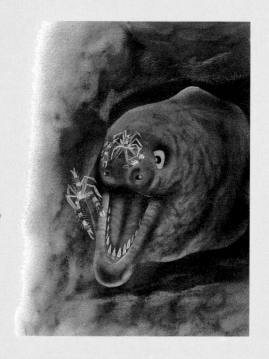

1. How many different species of crustaceans are there?
   a) About 30,000.
   b) About 154,000.
   c) About 10,000.

2. To strengthen their shells, crustaceans need:
   a) sodium.
   b) chloride.
   c) calcium.

3. The process in which the crab discards its old shell to grow a new one is called:
   a) captosis.
   b) caparcripsis.
   c) molting.

4. The largest existing crustacean is the:
   a) gigantic Chinese shrimp.
   b) giant Japanese spider crab.
   c) Philippine lobster.

5. The buccal pieces are:
   a) tail filaments.
   b) structures that help chew up the lobster's food.
   c) small stalks that support the crab's compound eyes.

6. Crabs are:
   a) carnivores.
   b) herbivores.
   c) omnivores.

The answers to CRUSTACEAN SECRETS questions are on page 32.

GLOSSARY

adapt: to make changes or adjustments in order to survive in a changing environment.

amphibians: cold-blooded animals that live both in water and on land. Frogs, toads, and salamanders are amphibians.

anemones: invertebrate animals that live in the ocean. Sea anemones usually attach themselves to rocks and shells and feed on plankton that they catch with their tentacles.

antennae: long, thin, paired, and jointed whiplike structures in crustaceans and insects that are used to detect anything in the nearby environment.

aquatic: of or relating to water; living or growing in water.

armor: a protective covering.

barnacles: small, hard-shelled sea creatures that fasten themselves to the bottom of ships or rocks.

brackish: slightly salty.

buccal pieces: relating to or near the mouth. In a lobster, for example, the buccal pieces are used for chewing.

calcium: a silvery white chemical element found in substances such as milk, bones, and shells.

cephalothorax: the united head and thorax (chest) of an insect or crustacean.

chitin: a rigid substance that forms part of the hard, outer shell of crustaceans and insects.

class: according to the scientific classification system, a class is the main subdivision of the phylum. For example, the class Crustacea belongs to the phylum Arthropoda.

compound eye: a type of eye found in crustaceans and insects that is made up of many simple eyes.

coral: a tiny, soft-bodied, salt-water animal that produces a hard, stony, cup-shaped shelter for protection. The coral stone built up by large colonies can form reefs and islands.

crustaceans: creatures with segmented bodies and an

exoskeleton, or shell. Lobsters and crabs are crustaceans.

current: a flowing mass of water.

epidermis: the outer layer of skin.

excrete: to get rid of, or discharge.

extremities: limbs of the body.

filaments: long, thin structures.

fossils: traces or remains of plants and animals found in rock.

fragment: a piece of something; a small part that is broken off from a whole.

gills: the structure inside crustaceans, fish, and many aquatic animals that removes oxygen from water for the animal's use in breathing.

habitat: the natural home of a plant or animal.

hepatopancreas: a gland that acts as both liver and pancreas in digesting and processing food.

larva: the wingless, wormlike form of a newly-hatched insect.

mate: to join together (animals) to produce young.

mollusks: animals such as clams and snails that usually live in water and have hard shells.

molt: to shed an outer covering or skin.

omnivores: animals that eat both meat and plants.

parasite: a plant or animal that lives on another plant or animal and gets its food from that plant or animal, called a host.

phylum (pl. **phyla**): according to the scientific classification system, a phylum is one of the main divisions of a kingdom. For example, crustaceans belong to the phylum Arthropoda, which is a main division of the animal kingdom.

plankton: tiny plants and animals that drift in the ocean.

predators: animals that kill and eat other animals.

prey: animals that are hunted and killed for food by other animals.

remains: what is left after something has died.

reptiles: cold-blooded animals

that have hornlike or scaly skin. Snakes, turtles, and lizards are reptiles.

species: animals or plants that are closely related and often similar in behavior or appearance. Members of the same species can breed together.

sperm: male reproductive cell.

statocyst: an organ of equilibrium, or balance, present in invertebrate animals. Lobsters, for example, have a statocyst at the base of each feeler.

tentacles: flexible, armlike body parts that certain animals use for moving around and catching prey.

transparent: allowing light to pass through so that objects on the other side or the inside can be seen clearly.

victim: one that is mistreated, injured, or killed.

ACTIVITIES

◆ Crab, lobster, and shrimp are popular foods. Visit the library and find some books about fishing for these animals. Where are some of these crustaceans caught? What are the methods used to catch them? Are there certain seasons for catching them? Do you think commercial fishermen should catch as many of these crustaceans as they can? Why or why not? Does water pollution affect these important food resources? If so, in what way?

◆ The horseshoe crab (sometimes called king crab) that lives along the Atlantic Coast of the United States is not really a crab. Look in library books or visit a display in a natural history museum to find out more about this animal. Can you think of why the horseshoe crab is often called a "living fossil?" What other creatures are its closest relatives? How is the horseshoe crab similar to and different from real crabs?

MORE BOOKS TO READ

The Crab on the Seashore. Jennifer Coldrey (Gareth Stevens)
Crabs. Christine Butterworth and Donna Bailey (Raintree Steck-Vaughn)
Crabs. Sylvia A. Johnson (Lerner)
Crabs. Kathleen Pohl (Raintree Steck-Vaughn)
Grasper: A Young Crab's Discovery Out of His Shell. Paul O. Lewis
    (Beyond Words Publications)
Harry Horseshoe Crab: A Tale of Crawly Creatures. Suzanne Tate
    (Nag's Head Art)
Hermit Crabs. Kathleen Pohl (Raintree Steck-Vaughn)
Looking for Crabs. Bruce Whately (HarperCollins)
Starfish, Seashells, and Crabs. George S. Fichter (Western Publications)
The World of Crabs. Jennifer Coldrey (Gareth Stevens)

VIDEOS

The Crab. Animal Families series. (Barr Films)
The Crayfish. Animal Families series. (Barr Films)
Crustacea. Oceans Alive series. (Environmental Media Corporation)
Crustaceans. (Encyclopædia Britannica Educational Corporation)

PLACES TO VISIT

The Montreal Aquarium
La Ronde
Île Ste-Hélène
Montreal, Quebec
H3C 1A0

**Sea World on the Gold
    Coast**
Sea World Drive Spit
Surfers Paradise
Queensland, Australia
4217

Fort Worth Zoo
1989 Colonial Parkway
Fort Worth, TX 76109

The Aquarium
Marine Parade
Napier, New Zealand

Pacific Undersea Gardens
490 Belleville Street
Victoria, British Columbia
V8V 1X3

Sydney Aquarium
Wheat Road, Pier 26
Darling Harbour
Sydney, Australia

**Toledo Zoological
    Gardens**
2700 Broadway
Toledo, OH 43609

INDEX

Answers to CRUSTACEAN SECRETS questions:
1. a
2. c
3. c
4. b
5. b
6. c